WHAT DOES IT LOOK LIKE?

IT HAS HOOVES!

BY KATHLEEN CONNORS

Gareth Stevens
PUBLISHING

first concepts

Some animals
have hooves.
These are hard parts
on their feet.

Many farm animals have hooves.
Horses have hooves.

Pigs have hooves.

Sheep have hooves.

Goats have hooves.

Wild animals have hooves too.
Deer have hooves.

Moose have hooves.

Bison have hooves.

Rhinos have hooves.

Giraffes have hooves.

Which animal has hooves?

Please visit our website, www.garethstevens.com. For a free color catalog of all our high-quality books, call toll free 1-800-542-2595 or fax 1-877-542-2596.

Cataloging-in-Publication Data

Names: Connors, Kathleen.
Title: It has hooves! / Kathleen Connors.
Description: Buffalo, NY : Gareth Stevens Publishing, 2026. | Series: What does it look like?
Identifiers: ISBN 9781482470871 (pbk.) | ISBN 9781482470888 (library bound) | ISBN 9781482470895 (ebook)
Subjects: LCSH Hoofs--Juvenile literature.
Classification: LCC QL942.C747 2026 | DDC 591.47'9--dc23

Published in 2026 by
Gareth Stevens Publishing
2544 Clinton Street
Buffalo, NY 14224

Designer: Andrea Davison-Bartolotta
Editor: Kristen Nelson

Photo credits: Cover, p. 1 (background) Igor Link/Shutterstock.com; cover, pp. 1, (goat), 21 (dog) Eric Isselee/shutterstock.com; p. 3 (bottom) erremmo.com/Shutterstock.com; p. 3 (top left) Madison Muskopf/Shutterstock.com; p. 3 (top right) Janet Griffin/Shutterstock.com; p. 5 Graphics Ark/Shutterstock.com; p. 7 dewa syah putra/Shutterstock.com; p. 9 Ginger Design/Shutterstock.com; p. 11 Nimbus Works/Shutterstock.com; p. 13 Andrea Izzotti/Shutterstock.com; p. 15 Maria 81/Shutterstock.com; p. 17 EB Adventure Photography/Shutterstock.com; p. 19 Tyrone Winfield/Shutterstock.com; p. 21 mariait/Shutterstock.com; p. 23 (bird) Sanit Fuangnakhon/Shutterstock.com; p. 23 (pig) Olhastock/Shutterstock.com; p. 23 (snake) Lana Langlois/Shutterstock.com;

Printed in China

CPSIA compliance information: Batch #QSGS26: For further information contact Gareth Stevens at 1-800-542-2595.